ARE YOU A TECHNOLOGIST?

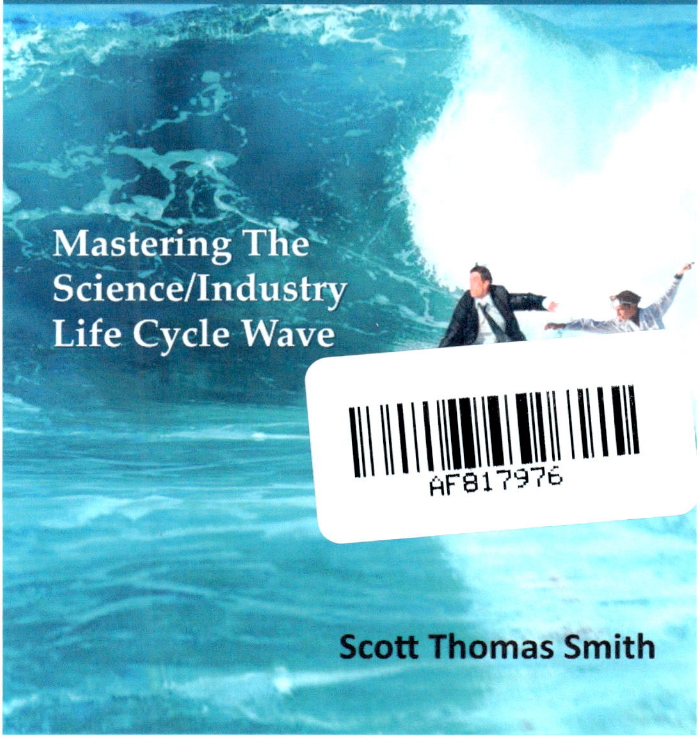

Mastering The Science/Industry Life Cycle Wave

Scott Thomas Smith

Did you know that everyone is riding the Science/Industry Life Cycle Wave? From the mobile device in your hand to the future of machine learning, self-driving vehicles and artificial intelligence; technology adoption is accelerating.

Businesses, organizations and non-profits are starving for the skillset of the Technologist. *Are You a Technologist?* will teach you to master the Science/Industry Life Cycle Wave. So, if your life's work is technical in nature or you are entering a new technology field, *Are You a Technologist?* is a must read to position yourself for the future.

Are You a Technologist? is delivered in a "Nano book" format (quick, actionable and to the point).

Visit Aperturance.com

ISBN 978-0-9840067-5-5

This book is dedicated to ….

All the Technologists that have shared their knowledge with me during my career. Thank you for investing in me.

Also, to my wife Marianne, the most creative person I know.

Contents

Are you a Technologist?

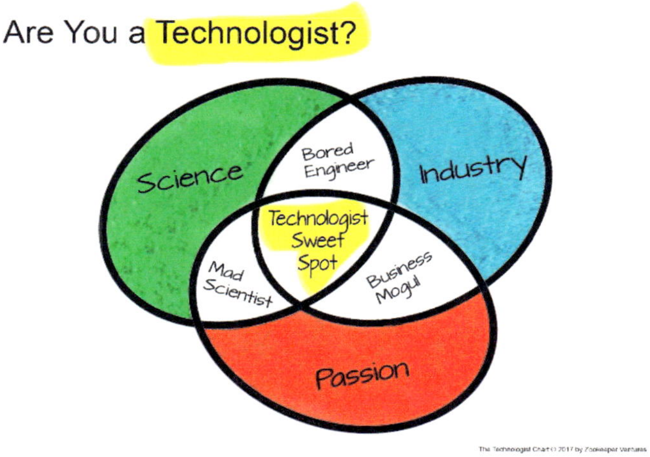

Do you spend your days at the server farm planting crops of petaflops or grinding away with machine learning trying to automate the automation, or maybe you are fresh out of collage with a shiny new diploma in engineering or the sciences? The Technologist Diagram is a simple chart to help you know if you are a true technologist living in the sweet spot between science, industry and passion.

The most basic definition of technology is, "the application of science to industry to solve problems." This definition forms the foundation of the technologist diagram. If you have a technical background, you understand the countless hours required to become an expert of the sciences.

However, you may not spend enough time considering the industry you serve, the company you work for or your own career strategy. Also, "Passion" is a subjective term and sometimes hard for the scientific mind to quantify. One way to describe passion is your skills, talents, preparation and attitudes combined to perform at your highest level. In other words, "Passion" is an investment. "Passion" is developed over time, and you bring "Passion" to your work every day.

A technologist living in the sweet spot is a highly-sought after and versatile person with the highest achievers moving through multiple proficiencies, positions, and companies in a career. I sometimes refer to the high achieving technologist as a Chameleon. Technologists can move between Science, Industry and Passion seamlessly and can adapt to nearly any situation just like a Chameleon.

Because the "Technologist Sweet Spot" is a crafted balance between Science, Industry and Passion, three things could happen when you are out of balance and outside the sweet spot.

- The Mad Scientist

  Science and Passion without Industry produces a Mad Scientist. Usually it's a penniless Mad Scientist because the market for their invention hasn't materialized. The Mad Scientist gets into the sweet spot by realizing business and industry are important and should not be trivialized.

- The Business Mogul

  Industry and Passion without Science generates a Business Mogul. The Business Mogul may be successful for a season, but without incorporating science into their industry, failure is inevitable as competitors advance. The Business Mogul moves to the sweet spot by embracing new science and not being intimidated by it. Remember, the Mad Scientist is building a robot to replace the Business Mogul so it's important to stay current with science.

- The Bored Engineer

  Science and Industry without Passion creates a Bored Engineer. A Bored Engineer will not change the world but by adding some passion they just might. A Bored Engineer can find the sweet spot by developing new skills, beefing up preparation and improving their attitude. Before long the Passion kicks in, the flow state happens and they are performing in the zone.

With our focus on the Sweet Spot, I will introduce you to the Science/Industry Life Cycle. This concept will illustrate how to position yourself for maximum impact as a technologist.

The Science/Industry Life Cycle Wave

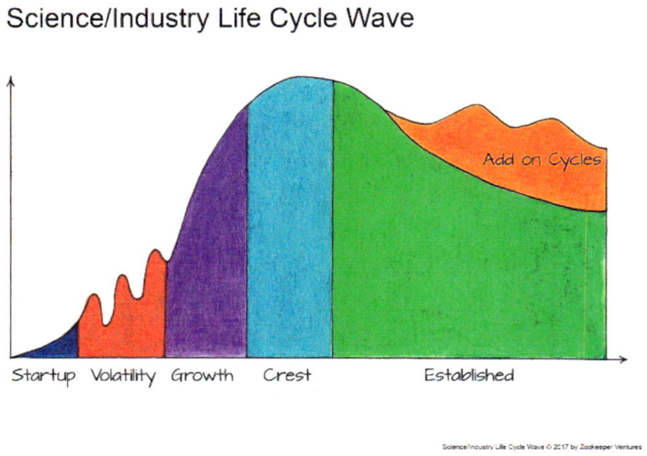

Have you ever looked at old photos of technology companies showing the employees? One of the most famous is the Microsoft photo showing a very young Bill Gates. We all know what happened to Bill Gates, but do you ever consider what happened to everyone else?

Early in my career I was in a company photo with about 30 of my co-workers. Over time It occurred to me how many had left the company. Out of curiosity, I started marking the photo every time one of my co-workers left. Before long only two of the original 30 were not marked on the photo. Then one day, I put the mark on me! My experiment with the photo was over and my days with the company were complete.

My 30 co-workers were all very talented and I consider all of them my friends. The company was growing and positive. New employees were being hired daily. I've always questioned what triggered people to leave and why I stayed longer. We were all riding the same Science/Industry Life Cycle Wave, but we all responded differently to the phases of the wave.

The Science/Industry Life Cycle Wave illustrates how a new science moves through an industry like a wave. The wave contains five phases including:

1. Startup

   The science is fresh and applied to an industry in small, controlled ways.

2. Volatility

   The science is gaining traction in the industry, but the short comings and limitations are appearing and must be overcome.

3. Growth

   The science adoption rate is high and the focus moves to implementation.

4. Crest

   The adoption of the science reaches saturation in the industry.

5. Established

   The application of the science in the industry declines and adoption slows. However, if the science can be applied to other industries, more "Add on Cycles" are possible.

If you work for a technology company or your life's work is technical in nature, then you are riding the Science/Industry Life Cycle Wave. Being complacent while riding the wave could cause you to be washed back to shore, washed up against the coral reef or washed out to sea drifting aimlessly. In the prior chapter, I introduced the Technologist Diagram and explained how the "Technologist" possess a crafted balance between Science, Industry and Passion and applies science to industry to solve problems. A Technologist is one of the few people who can navigate through each phase of the Science/Industry Life Cycle Wave.

My career experiences have allowed exploration of each phase of the Science/Industry Life Cycle Wave. I believe a well-rounded Technologist should experience each phase of the Wave. So, with our overview of the Science/Industry Life Cycle Wave in hand, we will deep dive into each phase of the wave.

The next chapter will describe the Startup phase, and examine the skill set required by the Technologist to influence and succeed in the Startup.

Startup Phase

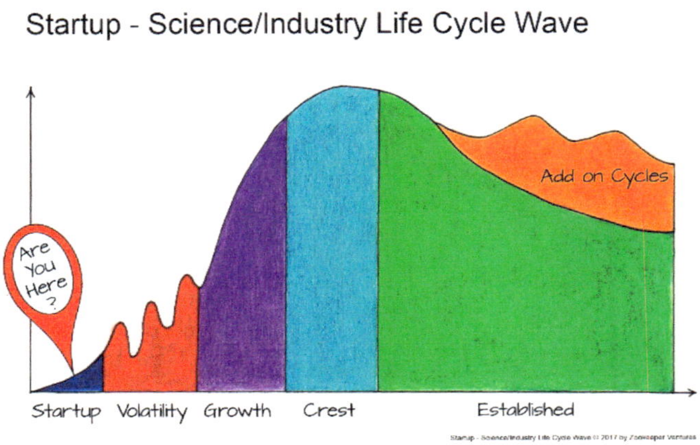

My first startup experience happened while working at the Oak Ridge National Laboratory in the office of Technology Transfer with my best friend Mike. Our role was to review the science developed at the National Lab and determine if it had application in industry. It was the dream job for a Technologist, and I lost count of the number of new science programs we reviewed and the market studies we developed. One program really captured our imagination. It was a new process called "CVDPCD" which stands for Chemical Vapor Deposited Poly-Crystalline Diamond. In other words, we were making diamonds!

So, Mike and I got busy with our startup idea. We applied the technologies developed at the Oak Ridge National Laboratory to the industry of diamond tipped cutting tools. Diamond tipped cutting tools are both expensive and complex to make and our startup idea solved those problems. We named our new startup Diamondback Industries.

A Startup making diamonds with a name like Diamondback Industries is a sure thing, right? Well, okay, we were not exactly making gem stones. We were growing thin diamond-like coating in a high temperature reactor. The science is called Carbide-Derived Carbon and is the root science for several developments including carbon nanotubes, graphene, and nanocrystalline diamonds.

We soon realized that perfecting our process would require considerable capital to mature the science. Our diamond reactor could grow loads of graphite, but was dreadful at growing nanocrystalline diamonds. In other words, the science was too early for commercialization. So, we decided to stop the program without additional losses. We learned many valuable lessons from Diamondback Industries, but the most valuable lesson was how to manage timing. Notice, I did not say how to manage "time". I said, "how to manage timing".

Successfully navigating the startup phase is not about luck; it's all about managing timing. It requires identifying and managing the right timing, not only for the science, but also the industry and even the problem being solved. To illustrate the point, let's compare the Startup Phase of two additional companies - sp3 Diamond Technologies and Diamond Foundry.

sp3 Diamond Technologies is a global leader in polycrystalline CVD diamond products and services. This company is a close analog to Diamondback Industries because the products and services are similar to our original Diamondback Industries startup business plan. However, sp3 Diamond Technologies started business 5 years after our Diamondback Industries startup. In other words, the science took approximately 5 years to mature and reach a point it could be commercialized. Diamondback Industries was 5 years early so we managed our timing and decided to stop. 5 year later, sp3 Diamond Technologies was on time, and managed their timing to start a viable long term company.

Diamond Foundry is growing synthetic gem quality diamond stones with a process perfected in the solar panel industry. Like Diamondback and sp3, Diamond Foundry uses a business model bypassing the traditional distribution channels and diamond cartel. However, what sets Diamond Foundry apart is the problem they are solving. Diamond Foundry is trying to rid the world of conflict diamonds aka "Blood Diamonds". Diamond Foundry's timing management for applying the science to the industry and identifying a high priority problem to solve is near perfection. By 2016 Diamond Foundry raised more than $100 Million in capital to fund the Startup Phase.

In the Technologist Diagram, I explained how the Technologist possess a crafted balance between Science, Industry and Passion. If you find yourself in the Startup Phase and manage your timing well, then bring the passion! Shifting your balance point toward passion will be contagious and amplify good timing. This also creates momentum to enter the Volatility phase of the Science/Industry Life Cycle Wave. The next chapter will examine why the Technologist is the perfect fit for the Volatility phase.

Volatility Phase

Volatility - Science/Industry Life Cycle Wave

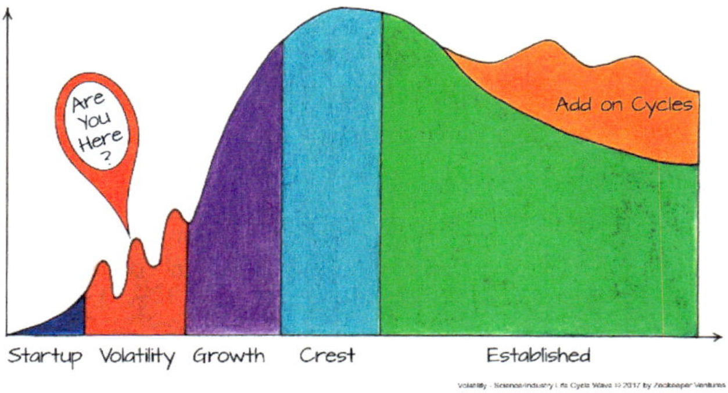

Volatility is my favorite phase of the Science/Industry Life Cycle Wave. In this phase of the wave, the science is gaining traction in the industry, but the short comings and limitations are appearing and must be overcome. The chart illustrates Volatility with an endless number of micro peaks and valleys. One day it's progress, and the next day it's problems. I'm not sure why I like this phase; it's challenging, hard to navigate and exhausting. However, this phase creates the knowledge and is the foundation for continued success through the Science/Industry Life Cycle Wave.

My first time through the Volatility phase was with American Magnetics, an early participant in the medical MRI market. My second time through the phase was with Elo TouchSystems, the company that invented the touchscreen. In both cases, leadership was a critical component to success. These are my 3 key leadership techniques for the Technologist in the Volatility Phase.

1. Balance = Leadership

   It's mandatory to stay in proper balance. Some days require an emphasis on science, technical or operational issues while other days require an emphasis on business, marketing and sales issues. A Technologist will gravitate to a leadership position because the Technologist is naturally balanced. During Volatility, a balanced Technologist is the best navigator to set overall strategy because they are a rock in the storm. Good navigation leads to high credibility. High credibility leads to high influence. Per John Maxwell, "Leadership is influence".

2. The 2-5-3 Priority Concept

   The 2-5-3 Priority Concept states for every 10 good ideas, 2 provide the good returns and benefits, 5 are average with slightly positive or slightly negative benefits and 3 are simply bad ideas with detriments. Clearly the goal is to identify, prioritize and implement the 2 good ideas and completely avoid the 3 bad ideas. However, have you considered the 5 average ideas? These ideas are just good enough to warrant extra time, extra attention and extra resources. I contend, the Technologist should avoid the "Average 5" just as fervently as the "Detrimental 3". The "Average 5" are nothing but passion vampires. They suck up the Technologist's time, energy and resources. Do not be fooled! Do not pursue the "Average 5" passion vampires. I value the 2-5-3 Priority Concept so much that I devote a chapter to it later in the book.

3. Iteration is the only Phase Hack

   When implementing science to industry, I don't think a short cut or hack exists to bypass the Volatility phase. However, driving iteration will get you through the phase faster. Consider my 2-5-3 Priority Concept again. Out of 10 "good" ideas, 3 of them are actually bad and 5 more are a waste of time. How do you know which ones should be avoided when they all look good at the beginning? Frankly, you don't know until you start the iteration process and identify the ideas with the highest positive correlation to your goal. The technologist that fast tracks the iterative process has the advantage.

Now that we have our 3 leadership techniques identified, here's the big secret revealed! The Science/Industry Life Cycle Wave is one big iteration process to reveal the truly good ideas. For every 10 good ideas entering the wave, the 3 worst ideas are washed out during the Startup Phase. The 5 average ideas are washed out during the Volatility Phase. The remaining 2 ideas are the finalists and have the privilege to advance to the Growth Phase of the wave.

A Technologist making it to the finals has proven characteristics. They manage timing well, they make their own passion, they lead using a balanced approach and their priorities are fueled by fast iteration. If you possess these characteristic, get ready for the ride of your life. You are about to catch the Growth Phase of the Science/Industry Life Cycle Wave.

Growth Phase

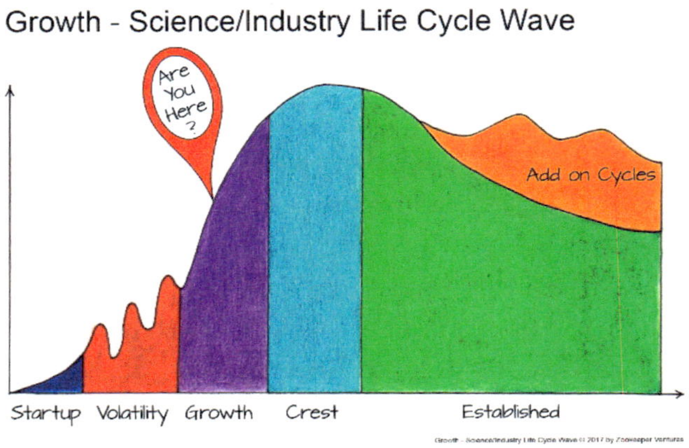

Mavericks, near Half Moon Bay in Northern California, has the world's most exciting surf breaks with waves reaching 50 feet or more. The Growth Phase is the "Mavericks" of the Science industry Life Cycle Wave. It's the phase that everyone dreams of catching. The Life Cycle Wave Chart illustrates the Growth Phase with a sharp, upward growth rate. In this phase, the science adoption rate is high, and the focus moves to implementation within an industry.

Let's recall how we arrived at the Growth Phase. In the last chapter, I revealed the big secret! The Science/Industry Life Cycle Wave is one big iteration process to reveal the truly good ideas. For every 10 good ideas entering the wave, the 3 worst ideas are washed out during the Startup Phase. The 5 average ideas are washed out during the Volatility Phase. The remaining 2 ideas are the finalists and have the privilege to advance to the Growth Phase of the wave. If you find yourself in the Growth Phase, be thankful because not everyone has the opportunity.

Being a technologist in the Growth phase is exhilarating. My first time through the Growth Phase was with American Magnetics, an early participant in the medical MRI market. My second time through was with Elo TouchSystems, the company that invented the touchscreen. Both times through the growth phase required a mastery of Momentum. These are my 4 key Momentum considerations for a Technologist in the Growth Phase.

1. Momentum: A Technologist's Secret Weapon

   Let's face it, an organization with momentum performs better than it usually would and an organization without momentum performs worse than it usually would. The same thing is true for a Technologist. A Technologist with momentum looks better than they truly are and without it they look worse than they truly are. Momentum can be the difference between success and failure. For example, look no further than the Falcon/Patriots Super Bowl 51 match up. The Falcons had all the momentum during the first half of the game. The Patriots had all the momentum in the second half. The carry over momentum from the second half propelled the Patriots to the first overtime victory in Super Bowl history. A small amount of momentum was the difference for the Patriots win. Like Tom Brady, use the power of momentum as your secret weapon.

2. Momentum is the Best Promoter of Disruption

   Many Technologists do not understand how to harness momentum to promote disruption in an industry. A Technologist may think they should stop momentum to make a change. However, we have all heard the saying, "It's hard to steer a parked car" or "It's hard to turn a motionless ship". Well, momentum is no different. Momentum is much easier to steer than to start. So, do not stop momentum. Harness it to promote disruption. A block under the drive wheels of a locomotive can keep it from moving, but a locomotive with momentum will crash through a thick concrete wall. The same is true for an organization with momentum. Disruption happens quickly and effortlessly as problems and obstacles are diminished with increasing momentum.

3. Understand Your Momentum "Makers"

   Momentum is fragile and most of the time we never understand the value of momentum until we lose it. Therefore, take some time to understand your momentum "makers". For a Technologist, momentum is almost always created by taking action in your area of strength. Actions that make you more confident, give you more energy, increase your expectations and strengthen your commitment are almost assuredly momentum makers. This is also important for an organization. An appreciation for momentum within an organization should be nurtured and guarded. A Technologist should be a momentum maker and should take the initiative to seek out other momentum makers in the organization.

4. Fear: Overcome the Momentum "Killer"

   The Growth Phase provides the adrenaline rush for the Technologist on the Science/Industry Life Cycle Wave. Most good adrenaline rush activities require overcoming fear. A Technologist that does not overcome fear will be a Momentum "Killer". Let's face it, the Technologist has devoted a lifetime preparing for the Growth Phase with much to lose and much to gain from how well they perform during the Phase. The pressure of the moment allows the fear of failure and sometimes the fear of success to gain a foot hold. To overcome the fear, the Technologist must rely on balance. Recall the Technologist Diagram. The Technologist possesses a crafted balance between Science, Industry and Passion.

> Balance is already a strength of the Technologist. Picture the Mavericks surfer riding a 50ft wave. The successful surfer is not out of balance. On the contrary, the surfer constantly adjusts the balance point to capture the momentum of the wave. So, overcome your fear and use your strength of balance to harness the full momentum of the wave.

Going through the Growth Phase is exhilarating. It's the reward for the Technologist's lifetime of preparation. Up next is the Crest Phase. We will consider the life changing decisions that will be made in the Crest Phase of the Science/Industry Life Cycle Wave.

Crest Phase

Crest - Science/Industry Life Cycle Wave

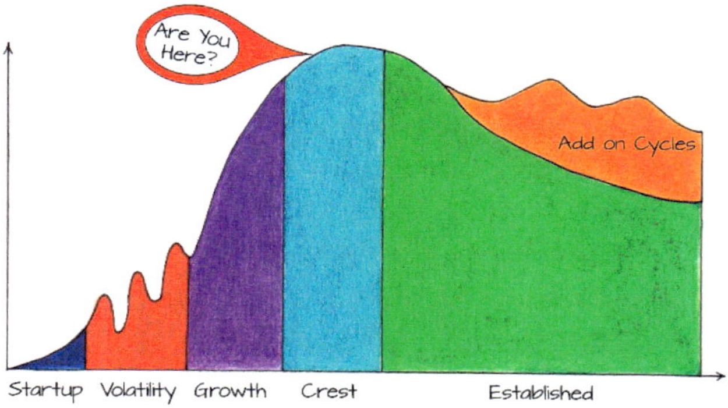

The summit, the pinnacle, the peak, being at the top; it all sounds good, doesn't it? The Crest of the Science/Industry Life Cycle Wave is the destination that every Technologist is vying to achieve. However, the Crest is not a true destination but a continuation of the wave. It's fluid and in motion like the other phases of the Wave. The Life Cycle Wave Chart illustrates the Crest Phase with a gradual slowing of momentum. This is the phase of the wave where change happens quickly and life changing decisions are made.

In many cases, the Crest Phase is the optimal time for companies and organizations to be sold or positioned for an initial public offering. The big financial returns for the Startup Phase investors and founders are typically made in the Crest Phase. This is also true for the Technologist. The point of highest professional value for the Technologist is at the top of the Crest Phase.

The success associated with the Crest Phase conceals the rapid change taking place in the market place as the science reaches saturation in the industry. The rapid change requires the Technologist to intentionally preplan key decisions early instead of making unintentional decisions at an inopportune moment. For example, a friend of mine was on a flight. During final approach for landing, the plane was caught in a wind shear. The plane was pushed down about 20 feet to the runway and bounced. The pilot immediately hit the throttle, pulled up, circled the airport and made another pass for a safe landing. My friend asked the pilot how he made the decision to put the plane back in the air so quickly, without thinking. The pilot explained this was a "preplanned decision". Pulling up and putting the plane back in the air during this situation was a decision he made 15 years ago while training in the Air Force. To prepare the technologist for the Crest Phase, these are my preplanned decision suggestions.

1. Decide to Grow Market Share

   The magnitude of the Crest Phase is determined by the market and how the science is accepted by the industry. The market acceptance does have an absolute limit in size. However, during the early part of the Crest Phase you and the organization possess the tailwind of momentum. Do everything to keep the momentum going and leverage the momentum to grow market share. Try to overestimate the importance of market share because capturing the last few percentage points of market share may be the difference between the market leader and all other competitors. The pilot in my example hit the throttle and pulled up because he had more space above the plane than below it. In the same way, a higher market share position provides more space to maneuver through the remainder of the Crest Phase, especially if you are the market share leader.

2. Decide to Stay or Decide to Go

Many Technologists do not understand and identify the optimal time to make big decisions. Let's face it, the worst time to make a decision is when you're in a rut, feeling depressed, can't see past the problems and only have mediocre options available. However, the top of the Crest Phase is the optimal time to make a life changing career decision. It's the point of highest professional value for the Technologist. It's the point at the top of the vista where you can see across the horizon, identify your options and choose the best path forward. Your intentional and preplanned decision may be to continue with your current organization and lead it into the Established Phase of the wave. You may have a favorite part of the wave and prefer to join a new organization that takes advantage of your strengths. The decision is yours, but the optimal time to make the decision is during the Crest Phase.

3. Decide Not to be Seduced by the Success of the Crest

   Sometimes the biggest deterrent of future success is today's success. We seem to think that past success will guarantee future success. It does not! Do not be seduced by the success of the Crest Phase. Did you ever play "King of the Mountain" as a kid? There was always that one kid who challenged your position on top of the dirt pile. Your competitors are no different. They are keen to knock you off the Crest. If you allow your ego to run rampant, feel entitled to your success or think you have "Arrived", get ready for a trip to the bottom of the dirt pile. Remember my definition of passion for a Technologist. Passion is your skills, talents, preparation and attitudes combined to perform at your highest level. In other words, "Passion" is an investment, developed over time, and you bring it to your work every day. If you focus on these things, you will not be seduced by the success of the Crest.

Going through the Crest Phase requires focus and planning. It's the best time to make career decisions. The next chapter will describe the Established Phase, and examine the skill set required by the Technologist to influence and succeed in the last Phase of the Science/Industry Life Cycle Wave.

Established Phase

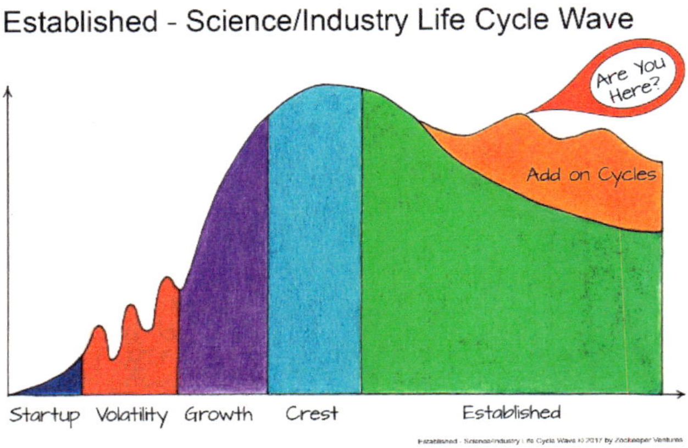

The name is Cow... Cash Cow. Are you really using a James Bond play on words to introduce the Established Phase of the Science Industry Life Cycle Wave? “Shocking. Positively shocking.” Well, it’s time we make the Established Phase of the wave cool again. The Life Cycle Wave Chart illustrates the Established Phase with a gradual decline to a steady level. This happens as the adoption of the science to an industry slows. This is the phase of the wave that tends to be stable with the possibility of “Add on Cycles” for continued success.

Many times, a Technologist will take a negative view of the old-school organizations in the Established Phase because they are incorrectly thinking the organizations are slow, bureaucratic and stagnant. However, the Science/Industry Life Cycle Wave is always in motion and moving. In the Life Cycle Wave Chart, one small detail is frequently overlooked. Notice the small arrowhead in the bottom right hand corner of the chart. That small arrowhead means the chart continues indefinitely in the Established Phase. The revenue and profits are sustaining and the Cash Cow is alive and well. After harpooning a bad guy, James Bond said, "I think he got the point". Get the point of the small arrowhead and don't allow yourself to overlook the organization in the Established Phase. The Established Phase represents a resource rich opportunity for the Technologist to pursue the next Add on Cycle of the Wave.

Let's take a closer look at the Add on Cycles. I propose that Add on Cycles are micro versions of the Science/Industry Life Cycle Wave imbedded within a larger wave. They are the "Mini Me" wave within the wave. (I couldn't resist a nod to Austin Powers) The scale of these "Add on Cycles" can be massive for the largest organizations. For example, GE is investing in new business segments such as "Digital" and "Additive Manufacturing". AT&T announced the $109 Billion acquisition of Time Warner to vertically integrate more content into the wireless business. I chose GE and AT&T as examples because GE started in 1889 and AT&T started in 1877. These companies have expertly managed Add on Cycles for over 100 years.

Google, the gold standard of Silicon Valley start-up culture, entered the Established Phase in 2016. For the first time, 50% of the global population has access to the internet. Out of the 7.4 Billion global population, 3.7 billion people are now online. When any market reaches the 50% saturation point the growth rates start to moderate. So yes, even Google has entered the Established Phase.

In 2015 Google reorganized into Alphabet. Alphabet is the umbrella organization with about 15 primary subsidiaries including Google, Nest, Waymo (self-driving cars) and Google X to name a few. Most interesting are the Moonshot projects primarily in the Google X subsidiary. Google's Moonshot projects are big, bold ideas that if successful could scale up to become huge Add on Cycles. In other words, Alphabet has created an incubator for Add on Cycles. Each Moonshot project is on its own "Mini Me" Science Industry Life Cycle Wave.

The 2-5-3 Priority Concept suggests that for every 10 good ideas entering the wave, the 3 worst ideas are washed out during the Startup Phase. The 5 average ideas are washed out during the Volatility Phase. The remaining 2 ideas are the finalists and have the privilege to advance to the Growth Phase of the wave. Even Alphabet Moonshot projects succumb to the Life Cycle Wave. Waymo, the self-driving car project, is promoted to become a subsidiary while other Moonshot projects are deemphasized.

For the Technologist, organizations in the Established Phase can provide career excitement and fulfilment like you might find in the Startup and Volatility Phases of the Live Cycle Wave without the downside risk. However, like any good James Bond motion picture, the Technologist needs to occasionally defeat a villain. In keeping with sinister villain names, let's call our villain Dr. Wrightsize. Notice the downslope in the early part of the Established Phase. Dr. Wrightsize shows up to reorganize and "right size" the organization to match current business levels and identify talent for the next Add on Cycle.

In *Octopussy*, the villain Kamal Khan comments to Bond, "You seem to have this nasty habit of surviving". Bond replies, "You know what they say about the fittest". In our quest to defeat Dr. Wrightsize, being the "Fittest" Technologist is good advice. A commitment to lifelong learning is important. Bringing new solutions together to solve problems by staying current on new science and up to date on industry trends is how the Technologist adds value to the organization and is in position to capitalize on the next Add on Cycle.

Going through the Established Phase brings us to the end of the Science/Industry Life Cycle Wave. Since we have a James Bond theme in this chapter, let's close it by considering if James Bond is a Technologist. The "Technologist Sweet Spot" is the crafted balance between Science, Industry and Passion. James Bond takes the science invented by

"Q", applies it to the industry of World Domination and solves problems associated with rouge super-villains. I'm sure Bond would say his commitment to passion in his work is unrivaled.

In *License To Kill*, Bond sums it up for Franz Sanchez:

> Bond: "In my business, you prepare for the unexpected."
>
> Franz Sanchez: "And what business is that?"
>
> Bond: "I help people with problems."
>
> Franz Sanchez: "Problem solver."
>
> Bond: "More of a problem eliminator."

Hmm, sounds like a Technologist to me.

Goldfinger. Honor Blackman, Sean Connery, Gert Fröbe, Guy Hamilton, Richard Maibaum, Twentieth Century Fox Home Entertainment. (2009).

Thunderball. Terence Young, Sean Connery, Kevin McClory, Ian Fleming, Jack Whittingham, MGM Home Entertainment, Twentieth Century Fox Home Entertainment. (2012).

Octopussy. Glen, J., Broccoli, A. R., Wilson, M. G., Fraser, G. M. D., Maibaum, R., Hume, A., In Davies, P., Twentieth Century Fox Home Entertainment Ltd. (2012).

License To Kill. John Glen, Timothy Dalton, Carey Lowell, Robert Davi, Talisa Soto, Twentieth Century Fox Home Entertainment, Inc. (2015).

Optimal Entry Points

Entry Points - Science/Industry Life Cycle Wave

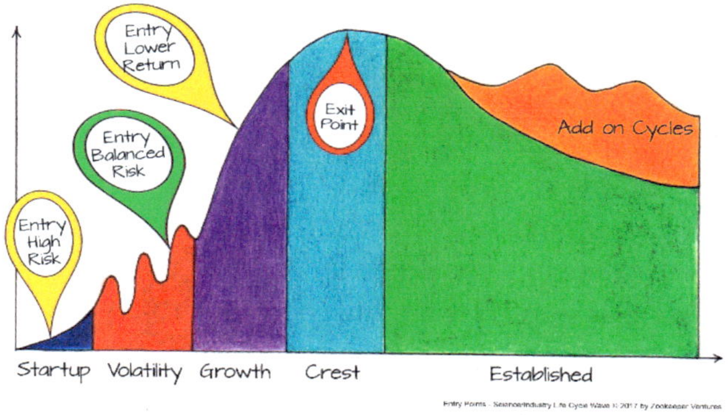

In the Crest Phase chapter, I identified the top of the Crest Phase as the time to make a life changing career decision. It's the optimal exit point because it's the point of highest professional value for the Technologist.

However, we are not always able to plan an exit with precision. Sometimes the exit may not be of our own decision. In the event of an unplanned exit, we need to prepare for the unexpected and have our next optimal entry points already considered and actionable. These are my "Red, Yellow and Green" optimal entry points:

- Red Crest

  The Crest Phase is the exit point not the entry point. An organization in the Crest Phase has limited upside potential.

- Yellow Growth

  Entering in the Growth Phase is easy and the path of least resistance. This is the phase where organizations will hire constantly. However, if you enter late in the Growth Phase, the Crest Phase is approaching quickly. Everything is great for the short term but then you hit the Crest and possible slowdown. You are disadvantaged because you have not had time to build up experience and a sense of intuition about positioning yourself.

- Yellow Startup

  Entering in the Startup Phase has the potential for huge upside and huge downside. Entering in the Startup Phase depends on your individual tolerance for risk. The 2-5-3 Priority Concept states that only 2 out of 10 startup ideas will provide nearly all the positive gains. That also means that 3 out of 10 will fail and 5 more will be average. On the other hand, the odds of winning the Publishers Clearing House Sweepstakes is 1 in 2.6 Billion. I prefer the odds of joining a Startup and working hard for success over the lottery odds any time.

- Green Volatility

  The Volatility Phase is my favorite time to enter the Science/Industry Life Cycle Wave. It is the proper balance between lower risk and capturing most of the upside growth potential. For example, let's assume most of the underperforming Startups have been washed out of the Wave and 4 organizations are left in the Volatility Phase. In all likelihood, these 4 organizations will have positive upside potential and you have a 50% chance that 2 of the organizations will have huge upside potential when they move into the Growth Phase. The Technologist also has an added benefit of learning the business during the Volatility Phase prior to entering the Growth Phase.

- Red, Yellow and Green Established

  The Established Phase is typically the lowest risk environment; however, the returns can be limited also. Recall Add on Cycles are the "Mini Me" or micro versions of the Science/Industry Life Cycle Wave imbedded within a larger wave. Therefore, the exit point would be at the crest of an Add on Cycle and the optimal entry points would be in the startup, volatility and growth parts of the Add on Cycle.

A Technologist is one of the few people who can navigate through each phase of the Science/Industry Life Cycle Wave. Improving your sense of timing and balance when choosing entry and exit points will keep your career momentum moving in a positive direction.

2-5-3 Priority Concept Illustration

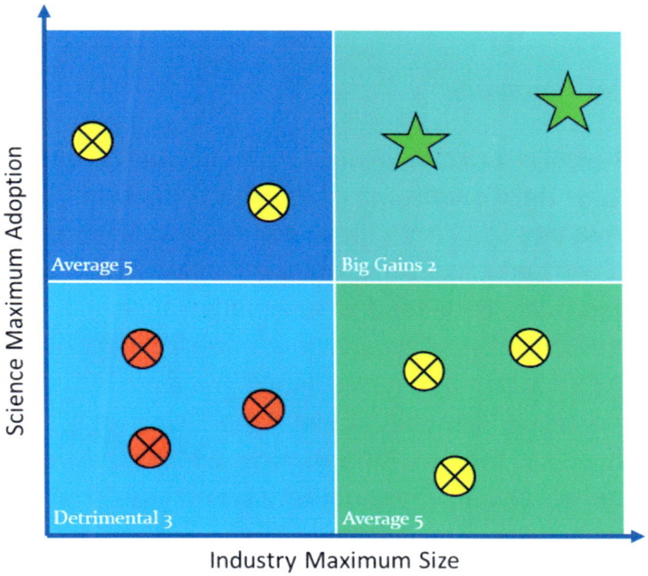

The chart above graphically illustrates the 2-5-3 Priority Concept for the adoption of a new science to an industry. Simply stated, for every 10 good ideas 2 provide the best returns and benefits, 5 are average and 3 are bad ideas with detriments.

- The Big Gains 2

  The two green stars in the upper right quadrant of the chart represent the "Big Gains 2". A high adoption of a new science in a large industry fuels the gains. These two ideas provide nearly all the positive returns from our ten initial good ideas. The "Big Gains 2" make it through the Growth Phase and into the Established Phase of the Science/Industry Life Cycle Wave.

- The Average 5

  The yellow circles located in the upper left and lower right quadrants of the chart represent the "Average 5". These ideas are just good enough to warrant extra time, extra attention and extra resources. They tend to have either high adoption of new science in a small industry or low adoption of a new science in a large industry. Remember, we should avoid the "Average 5". These are the passion vampires which steal your energy for only marginal returns on your investment. The "Average 5" are washed out during the Volatility Phase of the Science/Industry Life Cycle Wave.

- The Detrimental 3

  The red circles located in the lower left quadrant of the chart represent the "Detrimental 3". These are simply ideas that have negative results. The science has minimal adoption in a small industry. The Science/Industry Life Cycle Wave washes these out during the Startup Phase.

A real-world example of the 2-5-3 Priority Concept is the Gartner Hype Cycle chart for emerging science platforms. Gartner compresses more than 2,000 new ideas into a concise list of the science platforms with the greatest potential impact. It's the who's who list of science hype. Gartner publishes and updates the Hype Cycle chart every year. For reference, the 2016 version is shown below.

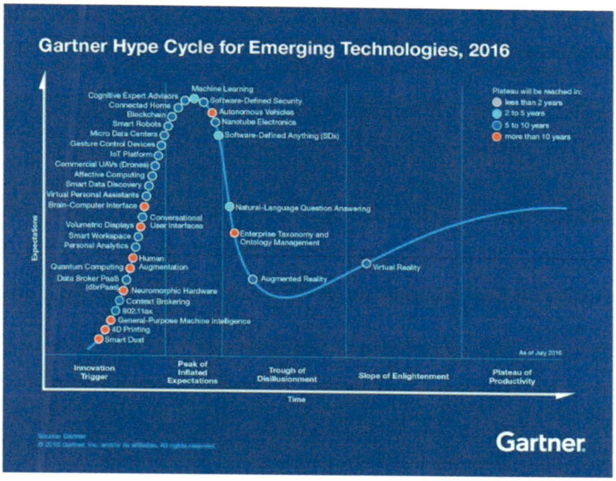

The Gartner Hype Cycle and the Science/Industry Life Cycle Wave do overlap. The science platforms listed on the Hype Cycle fall into the Startup and Volatility Phase of the Science/Industry Life Cycle Wave.

Let's use a 5-year period and compare the new sciences added to the Hype Cycle from 2011 through 2016 in the following table.

Gartner Hype Cycle New Entry Analysis		
Science Name	Year Entering Gartner Hype Cycle	Science/Industry Life Cycle Phase
3D Scanners	2012	Growth/Established
Crowdsourcing	2012	Growth/Established
Home/Mobile Health Monitoring	2012	Growth/Established
HTML5	2012	Growth/Established
Hybrid Cloud Computing	2012	Growth/Established
Mobile OTA Payments	2012	Growth/Established
Content Analytics/Recognition	2013	Growth/Established
Wearable User Interface	2013	Growth/Established
Cryptocurrencies	2014	Growth/Established
Digital Security	2014	Growth/Established
Autonomous Vehicles	2012	Startup/Volatility
Volumetric Display	2012	Startup/Volatility
Affective Computing	2013	Startup/Volatility
Quantified Self/Personal Analytics	2013	Startup/Volatility
Smart Dust	2013	Startup/Volatility
Virtual Reality	2013	Startup/Volatility
Connected Home	2014	Startup/Volatility
Smart/Cognative Advisors	2014	Startup/Volatility
Software Defined Anything	2014	Startup/Volatility
IoT Platform	2015	Startup/Volatility
Machine Learning	2015	Startup/Volatility
Micro Data Centers	2015	Startup/Volatility
Software Defined Security	2015	Startup/Volatility
802.11 ax	2016	Startup/Volatility
Blockchain	2016	Startup/Volatility
Context Brokering	2016	Startup/Volatility
Data Broker PaaS	2016	Startup/Volatility
Drones	2016	Startup/Volatility
Enterprise Taxonomy	2016	Startup/Volatility
Nanotube Electronics	2016	Startup/Volatility
Neuromorphic Hardware	2016	Startup/Volatility
Smart Data Discovery	2016	Startup/Volatility
Smart Workspace	2016	Startup/Volatility
Audio Mining/Bioacustic Sensing	2012	Washed Out
BYOD	2012	Washed Out
Complex Event Processing	2012	Washed Out
Consumer Telematics	2012	Washed Out
In Memory Analytics	2012	Washed Out
Silicon Anode Batteries	2012	Washed Out
Text Analytics	2012	Washed Out
Bio Chips	2013	Washed Out
Electro Vibration	2013	Washed Out
Neurobusiness	2013	Washed Out
Prescriptive Analytics	2013	Washed Out
Advanced Analytics With Self Service Delivery	2015	Washed Out
Citizen Data Science	2015	Washed Out
Digital Dexterity	2015	Washed Out
People Literate Technology	2015	Washed Out

We can identify the science platforms that have graduated to the Established Phase of the Science/Industry Life Cycle Wave, and we can identify the science platforms that were "Washed Out". We can also identify the science platforms that are still in the Volatility Phase and still listed on the Hype Cycle chart.

Gartner Hype Cycle New Entry Analysis Results	
Percent in the Growth/Established Phase	21%
Percent in the Starup/ Volatility Phase (Still on the Hype Cycle)	48%
Percent Washed Out	31%

According to the 2-5-3 Priority Concept, we should see approximately 20% of the new science platforms added to the Hype Cycle Chart reach the Growth and Established Phases of the Science/Industry Life Cycle Wave.

From 2011 through 2016 the Hype Cycle added 48 new science platforms. Ten (10) are identified as reaching the Growth and Established Phases. That's 21%. The correlation is good and confirms that 2 out of every 10 new good ideas provide most of the benefits.

Over the 5-year period, 23 science platforms (48%) entered and are still on the Hype Cycle list. All of these are in the Startup and Volatility Phases of the Science/Industry Life Cycle Wave. Science platforms entering in 2016 are still in the Startup Phase while others on the Hype Cycle list for multiple years are in the Volatility Phase. True to the Volatility Phase, some science platforms have moved ahead marginally while other have lost ground.

The 2-5-3 Priority Concept model suggests 50% of the new science platforms would be in the Startup and Volatility Phase. The actual result is 48%. Again, we have confirmation that 5 out of 10 new science platforms provide average returns; some positive and some negative.

As time moves forward, some science platform will graduate from the Volatility Phase and into the Growth Phase. These are the science platforms that overcome and resolve the limitations associated with applying a new science to an industry. Unfortunately, others are less successful and are washed out due to below average returns.

Several high-profile science platforms are still in the Volatility Phase including, Autonomous Vehicles, Virtual Reality, Machine Learning and Drones. It will be interesting to see which ones break out and make it to the Growth and Established Phases.

31% of the science platforms with a high level of Hype from 2012 through 2016 were washed out and have vanished from the Hype list. The 2-5-3 Priority Concept Model would have predicted 30%. This is a good correlation and proves the point that at least 3 out of 10 good ideas should be avoided.

The 2-5-3 Priority Concept works for characterizing large scale new science platforms over a long period of time and it also works for small, short term analysis. Some examples are, testing new ideas to improve a software application, evaluating ideas to improve a manufacturing process or reviewing new ideas to improve an on-line ad campaign.

As a Technologist, our primary responsibility is applying science to industry to solve problems. Choosing the best science options to apply to the best industries requires accurate decisions. Accurate decision making will become second nature, once we make the 2-5-3 Priority Concept a disciplined life habit. Remember, accurate decisions made today will have a positive compounding effect on each of our tomorrows as we move through the Science/Industry Life Cycle Wave.

Made in the USA
Monee, IL
05 November 2026

67383066R00029